Abdelhafid Mimouni

The Devastating Dual of Copper and Zinc

Abdelhafid Mimouni

The Devastating Dual of Copper and Zinc

ScienciaScripts

Imprint

Cover image: www.ingimage.com

This book is a translation from the original published under ISBN 978-620-6-70804-9.

Publisher:
Sciencia Scripts
is a trademark of
Dodo Books Indian Ocean Ltd. and OmniScriptum S.R.L publishing group

120 High Road, East Finchley, London, N2 9ED, United Kingdom
Str. Armeneasca 28/1, office 1, Chisinau MD-2012, Republic of Moldova, Europe
Printed at: see last page
ISBN: 978-620-8-03980-6

The Devastating Dual of Copper and Zinc

Author: Dr Abdelhafid Mimouni is an independent researcher specialising in bioinorganic systems chemistry, with extensive expertise in macromolecular synthesis and characterisation. He obtained his PhD in chemistry from the University of Paris XII in 1997 and a Diplôme d'Études Approfondies in bioinorganic systems from the University of Paris XI in 1993.

Summary:

This book takes an in-depth look at the imbalances between copper (Cu) and zinc (Zn), as well as metal poisoning, with a particular focus on mercury. By highlighting the sources of contamination, health effects, mechanisms of toxicity and detoxification strategies, it offers a holistic understanding of these crucial issues. Stressing the importance of ongoing research, it calls for a multidisciplinary approach to better understand the interactions between metals in the human body and to develop more effective prevention and treatment measures. This book thus provides an essential resource for health professionals, researchers and policy-makers involved in protecting public health from the dangers of heavy metals.

Table of Contents

Introduction

The balance between copper (Cu) and zinc (Zn) is a fundamental pillar of biochemistry and human health. These two trace elements are of crucial importance in a wide range of biological processes, from enzyme catalysis to protection against oxidative stress. Copper plays an essential role in the functioning of enzymes involved in cellular respiration, collagen formation and antioxidant defence, while zinc is an indispensable cofactor for more than 300 enzymes, notably involved in DNA synthesis, cell growth and immune signalling.

Precise regulation of the balance between copper and zinc is imperative. Inadequate levels or imbalances of these metals can lead to a multitude of pathologies, from metabolic disorders to neurodegenerative diseases. The complex mechanisms governing these interactions include transport proteins, hormonal regulation mechanisms and environmental interactions.

This introduction aims to highlight the crucial importance of maintaining an adequate balance between copper and zinc for human health. Excess copper, as seen in Wilson's disease, can lead to liver toxicity and neurological damage, while copper deficiency can lead to anaemia and bone disorders. Similarly, zinc deficiency can lead to growth retardation and immune deficiencies, while excess can disrupt copper absorption and induce mineral imbalance.

The interactions between copper and zinc are also influenced by other trace elements, notably selenium, which plays a crucial role in thyroid health and the detoxification of heavy metals such as lead (Pb) and arsenic (As).

This book sets out to explore in depth the biochemical mechanisms regulating the balance between copper and zinc, and their implications for human health. We will also examine the pathologies associated with these imbalances, such as Wilson's disease and liver poisoning, while highlighting the role of selenium in maintaining this balance and detoxifying heavy metals.

The specific aims of this book are manifold. We aim to clarify the biological roles of copper and zinc, analyse the pathologies linked to their imbalances, explore the functions of selenium, and propose therapeutic approaches based on recent research.

As a bioinorganist by training, my aim is to present this information in a clear and accessible way, while retaining the scientific rigour necessary for a specialist audience. This book is aimed at researchers, health professionals and anyone interested in the biochemistry of metals and their impact on human health.

Chapter 1: Biology and chemistry of copper and zinc

Copper (Cu) and zinc (Zn) are trace elements essential to life, playing diverse and crucial roles in numerous biological processes. Their presence and function in the human body are finely regulated to maintain homeostasis and prevent any imbalance that could damage health.

The biological role of copper : Copper exerts a significant influence in various biological functions. It is required as a cofactor for several enzymes, including cytochrome c oxidase, an essential component of the mitochondrial respiratory chain involved in ATP production. Copper is also involved in the biosynthesis of haemoglobin and neurotransmitters, regulates iron metabolism and promotes skin and hair pigmentation. In the central nervous system, copper is involved in the myelination of neurons and neuronal function, playing a crucial role in cognitive development. Copper is also necessary for corneal formation and ocular health, where it is involved in the synthesis of collagen and elastin, helping to maintain the structure and function of ocular tissues.

Biological role of zinc: Zinc is an essential trace element involved in a wide range of biological functions. As a cofactor for numerous enzymes, zinc is involved in processes such as the regulation of gene transcription, DNA replication and the stabilisation of cell membranes. It is also crucial for cell growth, differentiation and immune function. In the eyes, zinc is concentrated in the retina and crystalline lens, where it plays an essential role in visual function by participating in the synthesis of rhodopsin, the visual pigment required for night vision.

Metabolism and distribution in the human body: Copper and zinc are absorbed from the small intestine via specific transport mechanisms and are then distributed throughout the body via the bloodstream. The liver plays a central role in copper metabolism, while zinc is mainly stored in the muscles, bones, prostate and skin. The precise regulation of their intracellular concentration is ensured by specialised transport proteins and hormonal regulation mechanisms.

Interaction between copper and zinc: Although copper and zinc perform distinct biological functions, they share common metabolic pathways and can therefore interact within the body. These interactions can be competitive, cooperative or antagonistic, depending on the physiological context. For example, an excess of zinc may interfere with copper absorption, while a deficiency of zinc may lead to an increase in copper absorption to compensate.

Together, these aspects underline the crucial importance of copper and zinc in maintaining human health, as well as the complexity of their interactions within the body. An in-depth understanding of their biology and chemistry is essential for understanding potential imbalances and developing effective therapeutic strategies.

Chapter 2: Cu/Zn imbalances and associated pathologies

1. Definition and causes of Cu/Zn imbalances

Imbalances between copper (Cu) and zinc (Zn) in the body can result from a variety of factors, including unbalanced diets, metabolic disorders, problems with intestinal absorption, and even drug interactions. These imbalances disrupt the essential biochemical processes regulated by these metals, affecting the overall functioning of the body.

2. Health consequences

Imbalances between copper and zinc can have adverse health consequences because of their crucial role in many biological processes. Zinc is a cofactor in numerous metalloenzymes involved in essential enzymatic reactions, such as superoxide dismutase (SOD), which protects cells against oxidative damage, and carboanhydrase, which is necessary for acid-base balance. Copper, for its part, is necessary for the function of metalloenzymes such as cytochrome c oxidase, involved in the mitochondrial respiratory chain, and lysyl oxidase, which is crucial for collagen formation and tissue integrity.

3. Effects on the Immune System

Zinc plays an essential role in the optimal functioning of the immune system by regulating the proliferation and differentiation of immune cells, as well as the inflammatory response. It is also necessary for the activity of certain metalloenzymes involved in the immune response, such as thymulin, which regulates the maturation of T lymphocytes. An imbalance in the Cu/Zn ratio can compromise these immune functions, increasing the risk of infections and dysfunctional immune responses.

4. Neurological impact

Zinc is involved in neurotransmission and synaptic plasticity, while copper is necessary for the formation of neurotransmitters such as dopamine and noradrenaline. Imbalances between copper and zinc can disrupt these neurochemical processes, contributing to the development of neurological disorders such as Alzheimer's disease, Parkinson's disease and mood disorders.

5. Effects on skin and hair

Zinc is necessary for collagen synthesis, wound healing and regulating sebum secretion, making it a crucial element for skin health. It also helps regulate the proliferation of keratinocytes, the cells that make up the epidermis. Imbalances between copper and zinc can compromise these processes, leading to dermatological disorders such as acne, eczema and dermatitis. In addition, excess copper can disrupt the balance of melanocytes, leading to abnormal skin pigmentation.

Chapter 3: Wilson's disease and liver poisoning

Wilson's disease is an inherited disorder of copper metabolism characterised by excessive accumulation of copper in various tissues, particularly the liver and brain. The condition results from a mutation in the ATP7B gene, which codes for a protein involved in copper transport.

Pathophysiology of Wilson's disease: Mutation of the ATP7B gene leads to impaired copper transport in the liver, where copper is normally excreted in the bile. As a result, copper progressively accumulates in the liver, causing liver damage and increased release of copper into the bloodstream. Excess copper is then deposited in other organs such as the brain, kidneys and corneas, leading to dysfunction of these tissues.

Symptoms and diagnosis: The symptoms of Wilson's disease vary depending on the organ affected by the accumulation of copper. In the liver, symptoms may include hepatomegaly (enlarged liver), jaundice, ascites (accumulation of fluid in the abdomen) and abnormal liver function tests. Neurological symptoms include tremors, dystonia, movement disorders and personality changes. Diagnosis is based on blood tests to assess copper and ceruloplasmin levels, as well as genetic tests to detect mutations in the ATP7B gene.

Treatment and management: The treatment of Wilson's disease is aimed at reducing the accumulation of copper in the body and preventing associated complications. It is generally based on the administration of copper chelators, such as D-penicillamine or trientine, which bind to copper and promote its excretion by the kidneys. Low-copper diets may also be recommended. In severe cases, a liver transplant may be required to replace the failing liver.

Impact of excess copper on the liver: Accumulation of copper in the liver can lead to chronic inflammation, liver fibrosis and ultimately cirrhosis. Excess copper promotes the formation of free radicals, which damage liver cells and aggravate inflammation. Copper also interferes with lipid and carbohydrate metabolism, disrupting the liver's normal metabolic functions.

Cases of liver poisoning due to copper: Apart from Wilson's disease, liver poisoning due to copper can occur in the event of excessive ingestion of copper, often through sources such as food supplements, contaminated water or uncoated copper cookware. Symptoms include abdominal pain, nausea, vomiting, acute liver failure and, in severe cases, potentially fatal liver shock.

Chapter 4: The Role of Selenium in Health

Selenium, an essential trace element, is of vital importance in maintaining optimum health in humans. Its influence extends across various metabolic processes, particularly in thyroid function and its interactions with other minerals such as copper and zinc.

Selenium metabolism involves a series of enzymatic reactions that are crucial to the human body. These enzymes, such as glutathione peroxidase, play a vital role in protecting cells against oxidative damage, helping to reduce the risk of cardiovascular disease, cancer and neurodegenerative disorders.

More specifically, selenium is an essential component of thyroid hormones, in particular thyroxine (T4), a direct precursor of triiodothyronine (T3), which regulate basal metabolism and growth. As an enzymatic cofactor, selenium is essential for converting inactive T4 into active T3 in the thyroid gland, thus guaranteeing optimal hormonal balance.

Selenium also interacts closely with other trace elements such as copper and zinc. This interrelationship is crucial for maintaining mineral homeostasis in the body. Imbalances in these interactions can lead to a variety of pathologies, including thyroid dysfunction, changes in the immune system and metabolic disorders.

In short, selenium plays a central role in maintaining optimum health, acting on several levels to protect against disease and ensure the smooth functioning of various physiological processes.

Chapter 5: Maintaining the Cu/Zn balance using selenium

Selenium plays a crucial role in maintaining the balance between copper (Cu) and zinc (Zn) in the body, ensuring the proper functioning of numerous biochemical processes.

The biochemical mechanisms underlying this complex regulation mainly involve the action of selenoproteins, such as selenoprotein P and glutathione peroxidase, which act as mediators in the metabolism of copper and zinc. These proteins are involved in detoxifying heavy metals, modulating enzyme activity and protecting against oxidative damage.

Current studies and research highlight the importance of this delicate balance between copper, zinc and selenium for human health. Imbalances in these interactions can contribute to the development of diseases such as neurodegenerative disorders, cardiovascular disease and metabolic disorders.

The implications for health are vast and complex. An adequate intake of selenium, copper and zinc is essential to maintain mineral homeostasis and prevent nutritional deficiencies. Clinical studies have also shown that selenium supplements can improve health parameters in individuals with mineral imbalances.

Case studies provide concrete examples of the beneficial effects of selenium supplementation in maintaining Cu/Zn balance. Targeted nutritional interventions can help correct mineral imbalances and improve the overall health of individuals, offering promising prospects for the management of disorders associated with these complex mineral interactions.

Chapter 6: Detoxification of heavy metals

Detoxification of heavy metals is a vital process for human health, as these substances can lead to serious health problems if they accumulate in the body. Selenium plays a central role in this process, acting as a key element in the neutralisation and elimination of toxic metals.

Selenium exerts its detoxifying action mainly through its interactions with selenoprotein enzymes, such as glutathione peroxidase and thioredoxin reductase. These enzymes are involved in the detoxification of heavy metals by catalysing detoxification reactions, thereby reducing the oxidative damage caused by these toxic substances.

Selenium's mechanisms of action in the detoxification of heavy metals are complex and multifactorial. In addition to its antioxidant properties, selenium can form complexes with certain heavy metals, facilitating their excretion via the urinary or biliary tracts.

Extensive studies have examined the effects of selenium on the detoxification of lead (Pb), arsenic (As) and mercury (Hg), three heavy metals commonly associated with environmental health problems. These studies have shown that selenium can reduce the toxic effects of these metals by reducing their bioavailability and improving their excretion.

In addition to selenium, other elements also play an important role in the detoxification of heavy metals, including zinc, copper and sulphur. These elements interact in complex ways to neutralise and eliminate toxic metals from the body, underlining the importance of a balanced, nutrient-rich diet to support detoxification processes.

In conclusion, selenium is an essential element in the detoxification of heavy metals, acting as an antioxidant and facilitating the elimination of these toxic substances. Understanding the mechanisms underlying this action may help to develop strategies for preventing and treating diseases linked to exposure to heavy metals.

Chapter 7: Toxic effects of mercury (Hg)

Mercury, a chemical element that is ubiquitous in the environment, poses significant risks to human health when absorbed in excess. This section examines the different sources of mercury contamination, highlighting foods such as tuna and fish, as well as the permitted tolerance levels.

Overview of sources of mercury in the environment and foodstuffs

Mercury is found in the environment in various forms, including elemental, inorganic and organic mercury. The main sources of contamination include industrial emissions, electrical and electronic waste, and mining activities. However, a common source of dietary contamination is the organic mercury found in certain types of fish, such as tuna, shark and swordfish. These fish accumulate mercury because of their high position in the marine food chain.

Effects of mercury on human health, particularly on the nervous system, kidneys and neurological development.

Exposure to mercury can have devastating effects on human health, particularly on the central nervous system, kidneys and neurological development, especially in children and developing foetuses. Methylmercury, an organic form of mercury, is of particular concern because it can cross the blood-brain barrier and affect the brain. Symptoms of mercury poisoning include neurological disorders such as tremors, memory loss and learning difficulties.

Mechanisms of mercury toxicity in the human body

Mercury exerts its toxic effects by disrupting cellular processes and interfering with the enzymes and proteins essential for normal body function. It can also induce oxidative stress, damaging cells and tissues.

Possible interactions between mercury, copper, zinc and selenium

Complex interactions can occur between mercury and other elements such as copper, zinc and selenium. For example, selenium can have a protective effect by binding mercury and reducing its toxicity. However, high levels of mercury can deplete the body's selenium reserves.

Specific detoxification methods for mercury

Several detoxification methods can be used to reduce the mercury load in the body. These may include heavy metal chelators, sauna therapies and dietary changes to eliminate sources of mercury.

Case studies and practical examples of mercury poisoning and detoxification measures:

Case studies and concrete examples provide illustrations of the effects of mercury poisoning on human health, as well as detoxification strategies that can be implemented to mitigate these harmful effects. These cases can help raise public awareness of the dangers of mercury and encourage appropriate prevention and treatment practices.

Chapter 8: Therapeutic and preventive approaches

This chapter explores various therapeutic and preventive approaches to maintaining the balance of elements such as copper and zinc, as well as treating any imbalances and intoxications.

Strategies for maintaining the Cu/Zn balance

The balance between copper (Cu) and zinc (Zn) is crucial to the proper functioning of the human body. Strategies to maintain this balance can include dietary modifications, targeted supplements and management of environmental stressors that can disrupt this delicate relationship.

Selenium supplementation

Selenium is an essential mineral that plays an important role in protecting against the harmful effects of heavy metals such as mercury. Selenium supplementation can be an effective strategy for reducing mercury toxicity by promoting its binding and elimination from the body.

Diet and nutrition

A balanced, nutritious diet can make a significant contribution to preventing imbalances and poisoning. Diets rich in fruit, vegetables, whole grains and lean protein sources can provide the essential nutrients needed for the body to function properly and detoxify heavy metals.

Medical treatments for imbalances and poisoning

In cases of severe imbalance or proven intoxication, specific medical treatment may be required. This may include the use of heavy metal chelators to remove toxic substances from the body, as well as supportive therapies to alleviate symptoms and restore physiological balance.

Conclusion

-This book has explored in depth the various aspects of imbalances between copper (Cu) and zinc (Zn), as well as metal poisoning such as that caused by mercury. Summarising the key points covered, it is clear that understanding these phenomena is essential for the preservation of human health and the environment.

-Summary of key points

-We examined the sources of contamination, health effects, mechanisms of toxicity and therapeutic approaches for copper, zinc and mercury. Prevention, treatment and detoxification strategies were discussed in detail, highlighting the importance of a holistic approach to managing these problems.

-The importance of ongoing research

-Ongoing research in this area is crucial to furthering our understanding of the complex interactions between metals in the human body, as well as developing more accurate detection methods and more effective treatments. In addition, it is essential to conduct studies into the long-term effects of exposure to heavy metals and the best preventive practices.

-Future prospects for managing Cu/Zn imbalances and metal poisoning

Future prospects for the management of copper-zinc imbalances and metal poisoning will depend on a multidisciplinary approach involving collaboration between health professionals, researchers, policy-makers and the general public. Concerted efforts are needed to promote public health policies aimed at reducing environmental exposure to heavy metals, and to raise awareness of the risks and best practices for prevention.

-In conclusion, this book highlights the critical importance of maintaining a healthy balance between metals in our environment and our bodies, while underlining the need for ongoing research and concerted action to protect the health and well-being of all.

Glossary :

-Hereditary: genetically transmitted from one generation to the next.

-Protein: A molecule made up of amino acids that performs various functions in the body.

-Chelators: Substances capable of forming stable complexes with metal ions, making them soluble in water and facilitating their elimination by the body.

-Hepatic fibrosis: excessive scarring of the liver due to prolonged inflammation.

-Free radicals: Unstable molecules containing one or more unpaired electrons, capable of damaging cells by reacting with other molecules.

-Trace element: An essential chemical element required by the body in small quantities.

-Cofactor: A non-protein substance required for enzymatic activity.

-Homeostasis: The body's ability to maintain a stable internal balance despite external fluctuations.

-Myelination: The process of forming myelin around nerve fibres to speed up the transmission of nerve impulses.

Rhodopsin: Light-sensitive visual pigment present in the rods of the retina, essential for low-light vision.

-Metalloenzyme: An enzyme which contains a metal ion in its active site, necessary for its catalytic activity.

-Inflammatory response: The immune system's reaction to infection or injury, characterised by increased vascular permeability, infiltration of immune cells and release of inflammatory mediators.

-Neurotransmission: The process by which neurons communicate with each other, involving the release of neurotransmitters from a pre-synaptic cell and their binding to specific receptors on a post-synaptic cell.

-Keratinocytes: The epithelial cells that make up the outer layer of the skin, producing keratin, a fibrous protein that gives the skin strength and impermeability.

-Trace element: Chemical element required by the body in very small quantities to ensure it functions properly.

-Enzyme: Protein which catalyses chemical reactions in the body.

-Homeostasis: The body's internal physiological balance.

-Thyroid dysfunction: Impaired functioning of the thyroid gland.

-Triiodothyronine (T3) : Active thyroid hormone playing a crucial role in energy metabolism.

-Thyroxine (T4) : Inactive thyroid hormone, precursor of T3

-Bioavailability: Capacity of a substance to be absorbed and used by the body.

-Excess: Process of eliminating undesirable or toxic substances from the body.

-Antioxidant : Substance that protects cells against damage caused by free radicals.

-Heavy metal chelators: Substances that bind to heavy metals in the body, facilitating their elimination.

-Supplementation: The administration of nutrients in the form of food supplements to compensate for nutritional deficiencies or support health.

-Nutritional imbalance: A condition in which the levels of essential nutrients in the body are disturbed, which can lead to health problems.

-Detoxification: The process of eliminating toxins and harmful substances from the body, often carried out by the liver and kidneys.

-Methylmercury: A highly toxic organic form of mercury, often found in predatory fish such as tuna and swordfish.

-Heavy metal chelators: Substances that bind to heavy metals in the body, making them easier to excrete.

-Oxidative stress: An imbalance between the production of free radicals and the body's ability to neutralise their harmful effects, which can lead to cell damage.

-Blood-brain barrier: A physiological barrier that regulates the passage of substances between the bloodstream and the brain, protecting the brain from toxic substances.

References :

Vallee BL, Falchuk KH. The biochemical basis of zinc physiology. Physiol Rev. 1993;73(1):79-118.

Lutsenko S, Barnes NL, Bartee MY, Dmitriev OY. Function and regulation of human copper-transporting ATPases. Physiol Rev. 2007;87(3):1011-1046.

Cousins RJ, Liuzzi JP, Lichten LA. Mammalian zinc transport, trafficking, and signals. J Biol Chem. 2006;281(34):24085-24089.

Haldimann M, Boeckx RL, Flatz G. Zinc absorption in humans: a kinetic model. Am J Physiol. 1995;269(2 Pt 1)

Prasad AS. Zinc in human health: effect of zinc on immune cells. Mol Med. 2008 May-Jun;14(5-6):353-7.

Uriu-Adams JY, Keen CL. Copper, oxidative stress, and human health. Mol Aspects Med. 2005 Aug-Oct;26(4-5):268-98.

Maret W. Zinc and human disease. Met Ions Life Sci. 2013;13:389-414.

Squitti R. Copper dysfunction in Alzheimer's disease: from meta-analysis of biochemical studies to new insight into genetics. J Trace Elem Med Biol. 2012 Mar;26(1):93-6.

Pal A, Prasad R. Zinc and its role in Parkinson's disease: A review. J Neurosci Res. 2020 Mar;98(3):554-562.

Roberts EA, Schilsky ML; American Association for Study of Liver Diseases (AASLD). Diagnosis and treatment of Wilson disease: an update. Hepatology. 2008;47(6):2089-111.

Bull PC, Thomas GR, Rommens JM, Forbes JR, Cox DW. The Wilson disease gene is a putative copper transporting P-type ATPase similar to the Menkes gene. Nat Genet. 1993;5(4):327-337.

Ala A, Walker AP, Ashkan K, Dooley JS, Schilsky ML. Wilson's disease. Lancet. 2007;369(9559):397-408.

Brewer GJ. Wilson's Disease: A Clinician's Guide to Recognition, Diagnosis, and Management. Springer Science & Business Media; 2011.

Rayman, M. P. (2000). The importance of selenium to human health. The Lancet, 356(9225), 233-241.

Schomburg, L. (2016). Selenium, selenoproteins and the thyroid gland: interactions in health and disease. Nature Reviews Endocrinology, 14(3), 160-171.

Khan, M. A., Wang, F., & Iqbal, M. Z. (2019). Interaction of heavy metals with selenium and its uptake mechanisms in plants. In Selenium in Plants (pp. 279-300) Springer, Cham.

Khan, M. A., & Khan, S. (2017). Selenium: A potential mitigator of arsenic toxicity. Archives of environmental contamination and toxicology, 73(1), 21-31.

Brewer, G. J. (2015). Copper excess, zinc deficiency, and cognition loss in Alzheimer's disease. BioFactors, 41(1), 1-7.

Yanik, M., & Vural, H. (2005). Evaluation of the trace element levels in patients with essential hypertension. The Journal of Trace Elements in Experimental Medicine, 18(3), 167-172.

Clarkson, T. W., Magos, L., & Myers, G. J. (2003). The toxicology of mercury-current exposures and clinical manifestations. New England Journal of Medicine, 349(18), 1731-1737.

Grandjean, P., & Landrigan, P. J. (2006). Developmental neurotoxicity of industrial chemicals. The Lancet, 368(9553), 2167-2178.

Hightower, J. M., Moore, D., & Castillo, M. (2006). Levels of heavy metals in hair as an index of body burden. Journal of Occupational and Environmental Medicine, 48(8), 772-779.

Printed by Books on Demand GmbH, Norderstedt / Germany